Nachos and Numbers

...a Step by Step Guide to Hosting a Successful School Family Math Event

by

Thomas John Ecker

authorHOUSE™

1663 LIBERTY DRIVE, SUITE 200
BLOOMINGTON, INDIANA 47403
(800) 839-8640
WWW.AUTHORHOUSE.COM

First published by AuthorHouse 05/05/05

ISBN: 1-4208-3959-4 (sc)

Printed in the United States of America
Bloomington, Indiana

This book is printed on acid-free paper.

Introduction

Since the 1980's, elementary schools across the nation have often made sincere attempts to involve parents in the processes of educating their children.

Those attempts were often <u>not</u> made at the prerogative of school administrators, but federally mandated through legislation. Documentation of parent involvement has become routine. Family Math Nights have even become common annual events. Unfortunately, these events have also often become dull and unimaginative.

Let's examine this common scenario. Children and their parents are often seated at tables according to grade level, served a cold hot dog as well as a diluted cup of lemonade, and then expected to enthusiastically engage in an activity designed to address a grade-level Mathematics objective. In the worst instances, that activity is a cutesy classroom worksheet!

By the end of the evening, parents are either conferencing with teachers, or perhaps visiting with one another about last night's soccer game. Children are often left unattended to roam the campus in wild "packs", while every teacher, parent, and administrator involved exits the evening with a colossal headache. See you again next year? Let's consider an alternative, shall we?

Introducing... Nachos and Numbers:

- a fun, cost-free family meal!
- lively Mariachi music, enhanced by festive South-of-The Border décor!
- a host of engaging Mathematics activities designed to address the elementary curriculum for multiple grade levels, often without the use of paper and pencil, with optimum participation by choice!
- rewards for student participation and parent involvement!

Allow me to guide you through an innovative way to engage parents, teachers, and children in the process of developing Mathematics concepts. It is <u>sure</u> to become an annual event your entire faculty, student body, and parent group will embrace for years to come!

Bien Venidos a Nachos y Numeros!

Table of Contents

Chapter 1

Aren't We Doing <u>Enough</u> Already?

The teaching of Mathematics has changed dramatically over the past two decades. Teachers, in most cases, continue to be appropriately trained, or retrained, to guide their students towards understanding numerical concepts, rather than procedurally having students learn by rote or repetition. It's been a wonderful journey that has inevitably led to a deeper understanding of Mathematical concepts for both children <u>and</u> teachers. The journey is <u>not</u> complete, however.

Along the path towards teaching meaningful Mathematics, parents, long-time 'partners in education', have been left uninformed. They have therefore been left unable to effectively help their children when needs arise. A recent study compiled by the Pew Hispanic Center and the Kaiser Family Foundation, released new details about how parents feel towards schools in general. Forty-five percent of Hispanic parents feel schools have improved in the past five years, compared to 31% of African-American parents, and only 25% of Anglo parents. Parent groups across the United States are asking schools to respectfully include them as true partners in education. "Schools need to engage parents and teach them how to teach their child," said Mercedes Alejandro, president of Parents for Public Schools in Houston, Texas, America's fourth largest city.

We, as educators and parents, have an obligation to address this growing need! Are we really "bridging the gap", or is the gap growing wider? Does public consensus suggest the latter? While situations vary, it would be wise to closely examine your own educational setting, and set forth a plan of action. In a collegial atmosphere of fun food, lively music, and engaging Mathematical activities, <u>Nachos and Numbers</u> is a solid plan of action to do just that.

Chapter 2

Securing Solid Support

Whether you're an administrator or educator seeking a 'fresh' new approach to involving parents, or perhaps a concerned parent involved in your school's Parent/Teacher Organization, achieving philosophical support for a project often equates itself with financial support.

You should likely begin securing support for this project through your school's principal, or administrator. The next level of support may be your local district school board, and after those levels of support are garnered, your school's PTA or PTO would be next. A well-designed, and entertaining presentation, available in PowerPoint format, is available on the Nachos and

Numbers website: www.nachosandnumbers.net or at the address listed below.

Nachos and Numbers

%Mr. Tom Ecker

1559 Quail Run Drive

Troy, Texas 76579

It's often been said, 'We are who we associate with.' Are you the person complaining that teachers and parents just don't seem to care these days? Or, are you the person, pro-active in nature, ready to seek out individuals who share a common vision for school improvement and reform? Upon sharing the ideals of this project, it should become clearer to you who you should avoid, and at the same time, which kindred souls among you are anxious to join you in this quest for shared learning.

Hosting Nachos and Numbers, a successful Family Math Event, will put you on the path to securing your school's reputation as a 'trusted center of learning'. That is an honor earned by few, yet desired by all institutions of learning. Are you ready to get started?

Chapter 3

Putting Your Event on Everyone's Agenda

Scheduling your Nachos and Numbers Event months in advance is a wise idea for various reasons you'll discover as a read this guide. Quite simply, setting a specific date on a calendar will get you started! Once you have a date and a 2½-hour time period established, it's time to plan strategically how you'll get the message out.

If you're a busy parent, when was the last time you cleaned out your child's backpack? If this is part of your daily routine, you're probably aware of every school function taking place. If you're like the typical parent, you often find that crumpled notice days or weeks after an event was scheduled and occurred. Worse yet, sometimes the notice didn't even leave the classroom and remains inside a student's desk. It's time to <u>STOP</u> sending home type-written notices with students. Listed below, you'll find more effective ways to ensure that attendance to your Nachos and Numbers Event far surpasses anyone's expectations:

1) Pass out notices directly to parents waiting outside your school building, or as cars pass through the pick-up line. This could also be affectively achieved through the morning drop-off as well. It's also a great opportunity to warmly greet parents!

2) Arrange for willing students with enthusiasm, confidence, and leadership capabilities to make daily announcements regarding your scheduled Nachos and Numbers Event over the school's public address system.

3) Take advantage of today's World Wide Web. Post announcements on your school's website, classroom teachers' websites, and well as through e-mails.

4) Ask your local radio station(s) to make regular public service announcements about your scheduled Nachos and Numbers Event. Here is an example: *Next Tuesday evening, Willowbrook Elementary will be hosting NACHOS and NUMBERS, a Family Math Event, from 5:00-7:30 PM. Free food, music, and games with prizes await all students and parents who attend, so <u>come join the FUN!</u>*

5) Most newspapers are more than happy to include articles about upcoming activities at local schools. It is important that you write down all details about the date, time, and location of your Nachos and Numbers event. It never hurts to deliver your news in person, rather than over the telephone. An appearance speaks for your

sincerity. Ask if the announcement can appear more than once, and remember to be cordial and appreciative.

6) Families seem to be eating out more frequently than ever. Post colorful fliers at all of your local pizza take-out restaurants, as well as other restaurants frequented by families from your school's neighborhoods. Take advantage of those highly visible areas.

7) Encourage classroom teachers to reward students who attend with a "No Homework Tonight" coupon.

You're Invited to:
**Nachos
And
Numbers...
A Family Math
Event!**
Tuesday, 2/17/04
5:00 to 7:30 PM
at Willowbrook
Elementary
Cost: **FREE**
Join the Fun!

Chapter 4

Just How Much Will This Cost?

Costs associated with hosting a Nachos and Numbers Event will depend greatly upon the support you've been able to secure through your school's administration, your Parent Teacher Organization, as well as the general economic status of the students and families your event will serve.

If your school has already made great strides in improving the teaching of Mathematics through the purchase of manipulatives, quality software, and meaningful Math materials, assembling the activities suggested in Chapters 10 and 11 will be relatively cost-free. If not, are you or someone you know, willing to write an Innovative Grant Proposal through a local education foundation? There are also grant applications available through

NCTM, the National Council of Teachers of Mathematics at the NCTM website. Keep in mind that funding is often available through local service groups within your community. Don't be afraid to ask! It is important to make your request in person through a presentation, or through a well-written letter on school stationary clearly stating the objectives of this project.

A detailed list of suggested grocery supplies appears in Chapter 6. Many large grocery stores or 'mart' stores are looking for schools they can adopt to support worthwhile education endeavors. I personally hosted a Nachos and Numbers Event using $375. in donated funds through a local HEB Grocery store.

Walmart is another example of a store that routinely supports local education endeavors. These same wonderful stores often have boxes of 'freebies' they're willing to share that can be used as children's rewards throughout your scheduled event. Many restaurants of all kinds are more than happy to give out coupons for free food items that can also be used for rewards. Just remember, it is a courteous and gracious act to mention the support of those businesses or support groups on your fliers or notices announcing your scheduled event.

Rewarding students for their attendance and time on task will perhaps be the single most costly element associated with hosting a Nachos and Numbers Event. However, rewards are a vital element towards motivating student and parent participation and involvement. Ideas for motivational rewards are addressed specifically in Chapter 9. Mail-order catalogs such as The Oriental Trading Company, Inc. specialize in large amounts of age-appropriate gifts and prizes that can be affordably purchased in large amounts. This is their website: www.orientaltrading.com

Two hundred dollars in purchased rewards will support an evening of Nachos and Numbers for hundreds of students and their parents.

Since the Mathematics activities outlined in Chapters 10 and 11 are designed to avoid paper and pencil activities, costs associated with printing or copying will be minimal. Any copying costs will likely be due to printing fliers and announcements. Quality and appearance are important!

Now that you realize hosting a Nachos and Numbers Event will not 'break' the school budget, it's time to move ahead with your plans. I want to remind you that your plans began with a calendar and a schedule.

Chapter 5

Help? I Can't Cook for 400 People!

TONIGHT'S
MENU
Nachos
Grande
and
Ice-Cold
Sodas
Enjoy!

Hosting any festive occasion for a large number of people can be overwhelming for many of us. The thoughts of running out of food or beverages with impatient people standing in long lines flooded my own imagination with fear! Yet, no one would equate Nachos Grande with real cooking, would they? Maybe...

Thomas John Ecker

Allow me to help relax your own fears, since in this short chapter I will clearly establish guidelines for you to follow to ensure that everyone attending your Nachos and Numbers Event will be content and well fed.

Speaking from experience, do not hesitate to schedule a meeting with the person in charge of your own school's cafeteria. These fine people are experts in their field, having to follow United States Department of Agriculture (Food and Service) guidelines most of us are not even aware of. Did you know that a cardboard tray of tortilla chips has to be a minimum serving size of 25 grams or 9/10 of an ounce to meet DA school lunch guidelines? Add to that a 1 oz. serving of ground beef, an ounce of nacho cheese, and ¼ cup of diced tomatoes as well as other garnishes. Involving Mathematics, this could turn out to be a lot of fun! How? Allow me to explain.

One of the most important strands of the elementary Mathematics curriculum is Measurement. By labeling all of the serving utensils such as a ½-cup cheese ladle, or teaspoons and tablespoons used to serve garnishes such as onions, green peppers, or jalapenos... applied learning is taking place. I am also suggesting that you include more than one SCALE on your serving buffet so that attendees can physically measure the total weight of their serving. Before you even have to suggest it, children and parents will be comparing the weights of their meals. Imagine the learning taking place! That is referred to as real-life application. Wouldn't it be wonderful if all learning could take place in those appropriately staged environments.

The following is a suggested grocery and supplies list to serve Nachos Grande and sodas/juice to approximately 400 guests. Again, consult your own cafeteria personnel for further advice and recommendations. This is merely a guide:

- Four #10 cans of Nacho Cheese (now available at most grocery stores as well as large grocery warehouses). These are rather large cans.

14

- Four boxes of White ROUND Tortilla Chips (each box containing three 2 lb. Bags of chips).
- 15-20 pounds of pre-cooked, taco-seasoned ground beef (ground turkey can be less expensive!).
- Four quart-size containers of diced tomatoes, green peppers, onions, and jalapeno peppers.
- Three quart-size containers of sour cream.
- Two pints of mild salsa (optional).
- Twenty 2-liter sodas: Cola, Diet Cola, Dr. Pepper, and Sprite (that's five of each) with plenty of ice available.

  Juice-based beverages and water could certainly be substituted for sodas, and it may be necessary to do so in some states with stringent nutrition guidelines.
- Cardboard Nacho Containers (available at a restaurant supplier or warehouse grocery).
- Beverage cups, spoons, and napkins.
- Four or more slow cookers, along with serving utensils (two for cheese, and two for seasoned ground beef).

Certainly, you should file your receipts, since unopened containers of products can usually be returned without question if you're ordering extra amounts of specific items. Talk to your grocery store's manager in advance. Receipts will also document accountability for depleted funds, and help with planning future events.

Slightly compulsive by nature, I've personally spent the evening before my own Nachos and Numbers Event browning twenty pounds of ground beef. I wouldn't recommend that to anyone else. There is likely a trustworthy parent in each of your school's classrooms who would very willingly purchase one pound of ground beef, brown and season it, and return it to school the day of your event. The same would be true for all of the other necessary garnishes and supplies. Perhaps you even have someone in mind who would willingly be in charge of this very important part of the Nachos and Numbers Event, delegating duties. They would surely deserve endless praise and adulation!

Chapter 6

Concrete Mathematical Experiences

It is widely accepted among Math educators worldwide that a Mathematics program for young children in the elementary grades should be firmly based in real-life application. Therefore, it is an educator's obligation to provide students with constant opportunities for concrete experiences that will eventually lead students to the pictorial and symbolic levels of understanding Math concepts. Those concrete experiences often do <u>not</u> include the use of paper and pencil. It is very important that children understand numerical concepts, versus procedural exercises that lack meaning and therefore do not lead to that deeper understanding.

Through the activities you'll assemble for your Nachos and Numbers Event, you will be helping parents to understand the value of concept development. Your own school's classrooms are likely full of either commercial or homemade Math manipulatives and equipment that encourage children to experience the world of numbers, shapes, and measurement. This doesn't suggest that those supplies are actually being used, but perhaps collecting dust on closet shelves. You're going to change all of that!

If for some reason you believe your school's supplies are regrettably lacking in the area of Math manipulatives and equipment, hosting Nachos and Numbers will set the proper tone to facilitate securing the proper supplies. It might involve approaching school administration, curriculum chairpersons, or even writing grants to secure funding. Whatever path you choose to begin your journey, you'll be making an integral difference in the lives of students learning the concepts of Mathematics!

Concrete Experiences Can Provide For a <u>SOLID</u> Foundation of Understanding

Chapter 7

What <u>Are</u> The Elementary Mathematics Objectives?

An elementary Mathematics curriculum is divided into parts, often referred to as strands. The NCTM (The National Council of Teachers of Mathematics) has established standards for each grade level that teachers and administrators should be well aware of. The strands are listed below along with a brief listing of the skills or objectives typically associated with that strand:

1. <u>NUMERATION</u>: similar/different, digit recognition, value recognition, comparing numbers, one-to-one correspondence, tally marks, counting on/backward, odd/even numbers, number patterns, sequencing numbers, numeral writing, place

value/expanded form, rounding and estimation.

2. <u>ADDITION</u>: the concept of, the symbol (+), sums to 5, 10, and 18, the commutative property, multiple digit addends, missing addends, problem solving, multi-digit addition with regrouping or trading.

3. <u>SUBTRACTION</u>: the concept of, the symbol (-), minuends to 5, and 10, missing subtrahends, the inverse of addition, problem solving, multi-digit subtraction with regrouping, trading, or breaking down.

4. <u>MULTIPLICATION</u>: recognizing it as repeated addition, the symbol (x), products to 9, 40, 100, and 144, patterns, area, arrays, problem solving, missing factors, multiplying by 10, 100, and 1000, multi-digit multiplication with regrouping, or trading.

5. <u>DIVISION</u>: the concept of, the symbol, the inverse of multiplication, problem solving, by multiples of 10, estimating quotients, estimating the divisor, multi-digit dividends, up to 3-digits divisors, as well as quotients.

6. <u>BASIC FACTS</u>: single digit addition, subtraction, multiplication or division facts often performed under time limits.

7. <u>FRACTIONS</u>: the concept of, physical models, shaded parts, partial sets, numerator and denominator, equivalent fractions, improper/proper fractions, mixed numbers, and all four basic operations involving fractions.

8. <u>DECIMALS</u>: the concept of, models, shaded regions, values of money, comparing, common fractions/ decimals, place value, estimation, and all four basic operations involving decimals.

9. <u>PERCENTS</u>: the concept of, manipulatives, shaded regions, ratios, the symbol (%), and decimal/

fraction/percent equivalents.

10. <u>GEOMETRY</u>: identifying 2-dimensional shapes, 3-dimensional shapes, symmetry, congruent, translation/rotation/reflections, tessellations, area, perimeter, volume, point, line, line segment, ray, parallel lines, intersecting lines, perpendicular lines, angles of all kinds/measuring and their identification, polygons, circles, and their properties.

11. <u>MEASUREMENT</u>: in standard and non-standard (metric) units, estimations with reasonableness, appropriate tools, units of length, weight, volume/capacity, temperature, time, the calendar, elapsed time, problem solving, the value of coins, counting and giving back change.

12. <u>STATISTICS/PROBABILITY</u>: similarities/differences, sorting/classifying by attributes, graphing, interpreting graph data, predicting outcomes, summarizing results, finding the mean, median, mode, and range of a given set of data.

13. <u>ALGEBRA</u>: patterns, tables of input/output, charts, graphs, ordered pairs, x/y coordinates

Before anyone PANICS in reaction to setting up activities designed to address all of the objectives listed, <u>RELAX</u>! You can't do it all in one event. Give careful consideration to these questions:

    a. Which strands are most important towards enhancing the Math curriculum of your school?

    b. Are there specific needs relevant to your school's success on your mandated state standardized tests?

    c. Do you have the support of colleagues willing to put together quality activities by adopting specific strands?

Visit with your school's Mathematics teachers to discuss the needs of your school and how hosting this event can help further parents' and children's understanding of Math concepts. These questions and subsequent discussions will be beneficial in establishing a real academic purpose for hosting a Nachos and Numbers event, beyond positive public relations.

In the next chapter, you will find at least three dozen activities suggested for your Nachos and Numbers Event, addressing the thirteen strands of the elementary Mathematics curriculum. Keep in mind that some activities actually address multiple strands. Also keep in mind that <u>less is sometimes more</u>! Along the way you will undoubtedly have ideas of your own that would better facilitate the development of these concepts. Are the activities you've created attractive, motivating, engaging, and educationally rewarding? If you're not convinced of that, simply ask classroom teachers to try them out with their own students. Then, make improvements and modifications as suggested by the teacher and his/her students.

Chapter 8

A Host of Engaging Activities

NACHOS AND NUMBER is designed so that parents and children can freely choose to participate in several activities throughout the evening, in a completely non-threatening atmosphere. In order to foster that independence, you will want to print and post the name of each activity with a brief, easy-to-read explanation (also in Spanish?), along with necessary materials on the tables, grouped together by strand. Trust me in knowing that parents will appreciate being informed about the value of these activities. The explanations I've mentioned can be very similar to the explanations

I've used to describe each activity listed below, prefaced with "Dear Parent".

Larger floor activities should be labeled and posted using classroom easels. Chapter 9 will help you manage the volunteers who will monitor the various activities outlined below.

NUMERATION:

▫ <u>Building With Linking Cubes</u>
These brightly colored cubes provide eye-hand coordination, fine-motor skills, and spatial awareness. It doesn't take children long to discover that a peg of one cube inserts into the hole of another. Most sets of cubes come with examples of items to be constructed using these shapes. They should be prominently displayed as a motivational tool. Many teachers have watched in fascination as even 4th, and 5th grade students were enthralled by building with linking cubes.

▫ <u>More or Less With Two-Color Counters</u>
Using decorated coffee cans with lids, place approximately 20 counters that are of different colors on each side. These can even be dried kidney beans that are spray-painted on one side. Provide directions for children to shake the can, dump out the contents, separate them by color, and then compare the amount of one color to the other using 'more' or 'less'. You can easily expand upon this objective by using 5x 7" index cards with the symbols < (less than), >(greater than), or = (equal to) used to compare the amounts, placed inside the cans.

▫ <u>Place Value Pocket Charts</u>
We utilize a base-ten counting system. Therefore 10 ones can be traded for 1 ten, and 10 tens

can be traded for 1 hundred. In Kindergarten, children are expected to be able to count from 1-100 before the end of the year. However, reading and comprehending the value of digits in larger numbers soon becomes a challenge without place value knowledge. Using construction paper and a stapler, have parents and students make labeled pockets like the one pictured below. Students will make their own sets of 0-9 digit cards to be placed in pockets. The important <u>value</u> of digits will become obvious to students when asked questions such as,

"What is the value of the digit in the hundred's place?" <u>400</u>

"What is ten more than 475?" <u>485</u>

Provide a sampling of questions for parents to ask their children using their place-value pocket charts?

Could they be made to include the hundred thousands and beyond? Absolutely! Encourage students in the upper elementary grades to add pockets to include decimal places, with a decimal point labeled "AND" that will enable students to correctly learn to read the decimal as a mixed number.

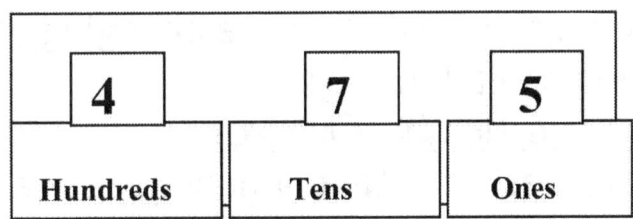

▫ <u>One to One Correspondence</u>

Very young students learn to count items by mentally attaching a specific number with a given amount or set as they count them. Provide students with brightly decorated containers of counters, such as bears, plastic chain links, or even plastic insects or worms. You'd be surprised how creative you might have to get in order to help nurture the development of one-to-one correspondence for some children. It can be lots of fun when you have a variety of resources.

▫ <u>Skip Counting With The Hundreds Board</u>

The hundreds board is an invaluable learning tool for students to internalize relationships between numbers. Using laminated hundreds charts or boards, have students and their parents use transparent counters (so that the numbers show through the counter) to skip count by 2's, 5's, 10's, etc. It can easily be used to teach before, between, and after as well. I suggest you purchase an inexpensive little activity book, with reproducible activity cards for non-commercial classroom use, from Learning Resources, Inc.. It is entitled <u>Hundred Number Boards</u>. It is filled with activities that address numeration issues such as 'more than', 'less than', and 'the same as'. They're engaging, entertaining, and even challenging!

▫ <u>The Nachos & Numbers Market</u>

This suggested activity encompasses a wide variety of Mathematics strands. It will also encompass a large area of floor space, but it's well worth every square foot! Start by sponsoring a non-perishable food drive in your school (or by collecting empty, but clean boxes, cans, and containers of grocery store food items). You will be donating the results of your food drive after your Nachos and Numbers Event. Price each food item with a number that can be 'rounded off' to the nearest dollar. For example, a box of cereal priced at $2.59 would be rounded to $3.00. Neatly arrange the food items on tables and shelves. Then, borrow plastic grocery baskets from your neighborhood grocery store. As 'customers' enter your store, they can be handed a basket, as well as a given amount of cash (play money in a plastic sealed bag) to spend. Instruct them to buy as many items as they can without going over the amount they were given. As students and parents pass through the check-out, the exact sum of their purchases (added on calculators) will be compared to the amount of cash in their bags. How close was their estimated vs. spent totals?

(NOTE: This particular idea has attracted <u>so</u> many participants in past events that we've had to quickly add volunteers to open up more check-out lanes to accommodate the long lines.)

▫ <u>Expanded Notation in Expanded Form</u>

Being able to identify the value of digits within a larger number is an important part of Numeration. Have construction paper pre-cut into 2" x 12" strips. With ready-made examples in clear view, students and parents will be asked to write numbers into the hundred thousands in expanded notation. Example:

300,000 + 50,000 + 2,000 + 800 + 90 + 6

If you then fold this paper strip five times, so that all 6 digits appear side by side, it will become apparent that it reads: three hundred fifty-two thousand, eight hundred ninety-six. Like an accordion, it can be stretched and read in expanded form.

ADDITION ACTIVITIES:

▫ That Nifty Little Number Line

1-20 number lines are an inexpensive tool for young children to learn to add or subtract small numbers. Their value towards developing an understanding of integers is undisputable. They are typically sold in rolls of 36, with adhesive backing to use on desk tops. Instruct parents to use the number line to solve a short page of addition sentences like this one: 6 + 8 =

By keeping the index finger of your left hand on six, use your right hand to 'hop' 8 more steps to the right on the number line. You should reach 14.

Suggestion: *Within a decorated wicker basket, place dozens of 35mm plastic film canisters, each containing a* 1-10 number line for Mom to keep in her purse, or simply to take home for future use. In a 'pinch' it can also be used as a tape measure!

▫ Base Ten Block Addition

In many elementary classrooms today, children are taught to add using base-ten blocks, or something similar such as Digi-Blocks. If you're not familiar with Digi-Blocks, explore their website to learn more:

www.digi-block.com

Transferring those experiences to paper involves 'bridging' to an abstract level of thinking. These concrete models provide the basis for a new vocabulary that includes terms such as 'trading', 'building', or 'packing'. These terms were relatively non-existent 15-20 years ago. By providing base-ten blocks in gallon-size plastic bags with ones (units), tens (longs), and hundreds (flats) available, allow students to share with their parents how they've been taught to add with these concrete models. Keep the number of problems down to a minimum... no more than five to ten problems. (NOTE: It is suggested that a friendly, knowledgeable teacher with base-ten block expertise, be on hand to assist students and their parents in this demonstration area, offering praise and encouragement.)

▫ Race For A Flat

A flat, in base-ten vocabulary, represents one hundred. A student and their parent can race against one another to be the first to trade 10 ones for a ten, and then 10 tens for a flat. Each player should be given one dice to roll, a supply of ones and tens, and a place value chart like the one below. The dice roll will determine how many units, or ones, they've earned in order to make trades. When one of the players has eventually traded 10 tens for a hundred or flat, a winner is declared!

TENS ONES

100!

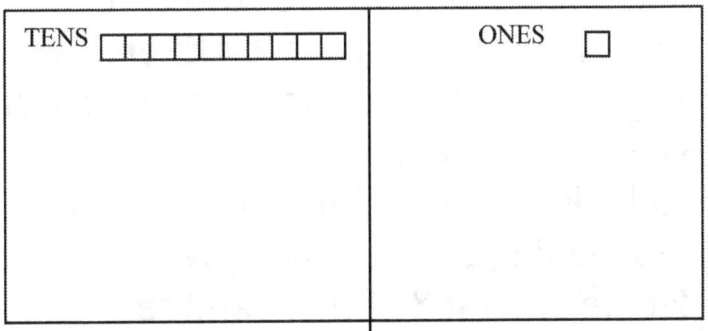

SUBTRACTION ACTIVITIES:

▫ <u>Problem Solving with Subtraction</u>

It is imperative that young children in the elementary grades understand the concept of an operation before an algorithm is ever introduced. By supplying parents with a short page of two word problems, and consumables to 'act out the problem', children will begin to learn the concept of subtraction as taking a given amount away. Write your own, or use the examples below along with the inexpensive consumables mentioned in the problems.

1. Jessica's mother gave her a small bag containing 12 animal crackers. As Jessica headed out the door, she reached into the bag and ate three crackers. How many animal crackers did Jessica have left in her bag?

2. Sam's teacher gave him a small bag of Hershey Mini-M & M's. She told him he could open up the bag and eat 10 of the candies, but he must give the rest of them to his mother or father. How many M & M's did Sam have left to give his Mom or Dad?

3. Sarah took five cheddar cheese crackers out of the box. She quickly ate 3 of the crackers. Her father then gave her 8 more crackers. Sarah decided to give him 3 crackers. How many crackers did Sarah have then?

4. Todd loves chocolate chips. When his mother was making cookies, she gave him 18 chips. Todd decided to eat 8 of them and save the rest to split

between his two friends. How many chocolate chips did Todd give to each of his friends?

▫ Subtraction Action

This is a fast-paced card game that is easy and fun to play! Two players should divide the deck equally with the cards facing down. Each player will place an overturned card in the middle of the table at the same time. The first person to produce the DIFFERENCE between the two card values gets to keep the cards. You will need to supply a page that lists the values of an Ace-1, Jack-11, Queen-12, and a King-13 for those people not accustomed to playing card games.

Consider purchasing one of many fine books with card games. Often times, an author will allow you to photocopy games for parents and kids as long as it done for non-commercial use.

▫ That Nifty Little Number Line

Like addition, the 1-20 number line can be a very useful tool in helping students to solve simple problems in subtraction. Provide parents/children with a short page of subtraction exercises within sentences like this one: 15 - 7 =

Placing your right index finger on 15, use your left hand to hop seven places or numbers to the left. You should arrive at 8.

▫ Subtraction With Base Ten Blocks

Providing laminated F-L-U boards (for flats, longs, and units), and reachable supplies of base ten blocks, provide students and parents with 5 -7 two-digit subtraction problems that involve regrouping or trading. Encourage students to share with their parents how they've been taught to subtract using base-ten blocks. With a knowledgeable, nurturing teacher as a resource, parents will understand why terms like 'borrowing' are no longer used in the subtraction algorithm.

Multiplication/Division Activities:

▫ Conceptually, it is imperative that students learn that multiplication and division are related operations. Using 2-color counters, centimeter cubes, or dried beans, students can explore these operations if provided with thought-provoking short stories that need solutions. I suggest you avoid calling them "problems". Who needs a problem, anyway, right? Dry-erase boards and markers will provide the 'bridging level' achieved through these stories. Here are a couple of examples:

1. *Jasmine and her mother are planting a garden. They want to plant 4 rows of tomatoes with 6 tomato plants in each row.*

   Using counters, can you illustrate how many tomato plants Jasmine and her mother will be planting.

   Then, using a dry-erase board and marker, write a multiplication sentence to represent what you did.

   Solution: 4 multiplied by 6= 24

2. Derrick has a collection of 28 toy cars. He wants to display them on four shelves in his bedroom. Using counters, can you figure out how many cars he should place on each shelf?

   Now, write a division sentence to represent what you did.

   Solution: 28 divided by 4= 7

▫ Roll An Array

Arrays are commonly used today to teach multiplication facts because they give facts meaning. They will also directly help a student to understand the Geometry concepts of Perimeter and Area. Classes will very often create entire booklets of arrays to represent all of their multiplication facts. Typically, 3rd graders will concentrate on the 1's, 2's, 5's, 3's, and 4's. Fourth graders will review those, and work to internalize the 6's, 7's, 8's, and 9's. By the fifth grade, most students are very capable of knowing all of their multiplication facts through the 12's. This simple game can get students started in understanding what multiplication facts are all about. A parent and child will need two die, and array paper divided into centimeter squares, then cut into rectangles similar to the one below.

Each player takes turns rolling the dice. Each person has their own rectangle. Using both numbers, color or shade in the area representing that fact. You could also use centimeter cubes or anything else to fill in the squares. For example, 3 x 4 = 12, or three rows of four squares in each row equals twelve. If someone rolls a product they can't shade on the rectangular array, they simply have to wait until the next time. The first person to shade in their entire rectangle wins the game.

Basic Facts Activities:

The following activities are all highly engaging, but require floor space to accommodate them. They'll be very popular among your Nachos and Numbers guests.

- ## The Addition Facts Cupcake Walk

  Using inexpensive flash cards, tape approximately one dozen of the cards in a large circle, on the floor. You could even use clear contact paper to tape down the cards because they're going to be stepped on throughout the evening. Put the answers, or sums, to the addition facts in a can. Using appropriate children's music, whenever the music stops, the walkers will stop, the caller will read a sum, and thereby declare a winner. It might be a good idea to post a large 1-20 number line on the wall or inside the circle for younger students who are just beginning to learn basic addition facts. Boxes with individually wrapped cupcakes make ideal, inexpensive prizes. Students will want to walk the circle over and over again, earning enough cupcakes for their entire family!

  Note: This activity can easily be adapted to a Multiplication Facts Cupcake Walk using Multiplication Flash Cards. Why not have two 'walks' running simultaneously, one for lower elementary grades, and one for upper grades?

- ## Bean Bag Subtraction Toss:

  On large brightly colored tag board, divide the boards into 9 equal parts. Like the examples below, one board will consist of larger numbers, and the other will consist of small numbers. Place the two tag boards side by side on the floor against a wall.

Players will stand behind a taped line, tossing one beanbag to a larger number, and one beanbag to a small number. Their job will be to subtract and produce the difference. Post a large number line above the activity for young students in the early stages of developing this concept.

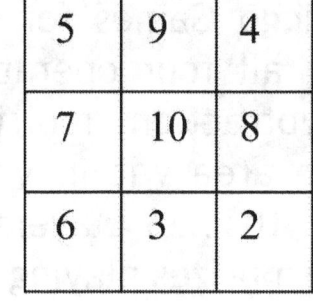

16	10	15
11	13	18
14	12	17

5	9	4
7	10	8
6	3	2

□ The Basic Facts ½ Liter Ring Toss

½ Liter sodas can be purchased in six packs. Arrange the sodas in a triangular arrangement similar to the one below. Each player will have the opportunity to answer three multiplication facts, if that student is in grades 3-6. For younger students, use number recognition cards, or basic addition/subtraction facts. If they answer all three correctly, they'll be handed three 4" rings (plastic bracelets work great!) to toss onto the soda bottles from behind a taped line. It would be wise to post an enlarged multiplication chart, hundreds chart, or a 1-20 number line above this activity so that students developing these concepts will be able to achieve success in answering their facts. If they 'ring' a

35

soda, it's theirs to keep. You may need to limit one per player.

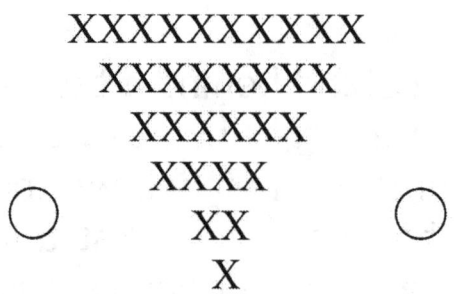

□ <u>Basic Facts Bingo:</u>

Bingo Games for up to 36 players are available in all four operations of Mathematics: addition, subtraction, multiplication, and division. Set up an area within your Nachos and Numbers Event so that 36 players can comfortably sit down and win prizes playing Bingo as a family. You may want to schedule Addition Bingo for the first hour, and Multiplication Bingo for the second hour. Provide zippered bags or plastic bowls of dried beans to use as game board markers/counters, as well as multiplication charts. Use an overhead projector and large screen to display the facts called, and under what letter they will be found. Prominently display the prizes winners will be able to choose from. Those prizes could range from calculators to kites, or from plants to pizza coupons!

Be prepared for attendees to race to the tables when you announce that Bingo is about to begin. For the very first time, the Special Education Teacher in my building met the parent of one of her students while playing Basic Facts Bingo at a Nachos and Numbers event. This particular parent had not attended even one parent/teacher conference, nor her son's annual Special Education Annual Review. It was clear that this parent was not comfortable in a formal education setting, but Nachos and Numbers was just the "ticket" she obviously needed to step through the doorway of her son's school! Who knows, after meeting her son's warm and friendly teachers, perhaps she'll be back.

Fractions, Decimals, and Percents

▫ <u>Thrifty Shopping at The Lost and Found Store:</u>
Is your Lost and Found Rack overflowing with unclaimed coats, jackets, hats, gloves, and a variety of other clothing items?
This will be a opportunity to put that clothing to use, while maybe even finding out where some of it belongs. First, start by organizing the clothing, hanging jackets together, coats together, hats and gloves together on a table, etc. Price each item with whole dollar amounts on tags large enough to also attach a colored round sticker. Why? You'll be providing shoppers with a sign something like this:

> **HUGE CLOTHING SALE**
> Yellow Sticker-**25% Off**
> Green Sticker- **50% Off**
> Red Sticker-**75% Off**
> *Grab a zip-lock bag of play money, and "Shop Until you Drop"!*

A friend in the retail clothing industry was often dismayed by the number of adult customers who did not have an understanding of common percents and fractions when shopping for sale items. Having a nurturing and knowledgeable adult (the Thrift Store Manager) on hand to help shoppers will probably be necessary. You will likely want to have calculators on hand for shoppers' use as well since this may be a "learning experience" for both children <u>and</u> adults.

▫ <u>Chocolaty Fraction Fun:</u>

This chocolate bar activity is an ideal way to introduce fractions meaningfully. By the way, having it located right next to your Thrift Store (and its Manager) will help prevent any chocolate bars from disappearing without students actually doing the activity. The following questions will need to be processed on handouts. A box of regular Hershey chocolate bars that are divided into sections is the only other thing you'll need. Enjoy!

Directions: Unwrap your chocolate bar, and carefully break the bar into sections along the lines. Lay those sections on a clean napkin or paper towel. **Do not be tempted to eat any of it while you are working because you will need all the parts to**

answer each question. You should be able to see that you have twelve equal sections, or parts. You are ready to begin!

1. A fraction is made up to two numbers, with one number written over another number. One-half (½) is known as a common fraction. The top number represents that you have 1 part out of 2 (bottom number) equal parts. If you ate just one section of your candy bar, what fraction of the candy bar did you eat? (Answer: 1/12 or one-twelfth)

2. What fraction of the candy bar would be left? (Answer: 11/12 or eleven-twelfths) Remember that there are still twelve equal parts in a complete bar.

3. You have decided that you want to share the candy bar with your Mom or Dad. What fraction of the candy bar will each of you get? (Answer: 6/12 or six twelfths) That's the same as ½ since six is half of twelve!

4. The top number of a fraction is called the numerator, and the bottom number is called the denominator. If your little brother ate five pieces of your chocolate, what fraction of the candy bar would you have left? (Answer: 7/12 or seven twelfths) Which number is the numerator? (7) Which number is the denominator? (12)

5. If you were so hungry that you ate all of the sections of your candy bar, what fraction would represent how many sections you would eat? (Answer: 12/12 or twelve-twelfths) When the numerator is equal to the denominator, that means one whole!

6. Now, arrange the twelve sections in four equal

groups. Do you have three in each group? If you gave one group, or ¼, to your school principal, what fraction names what you have left? (Answer: ¾ or three fourths) By now you realize that your fraction's denominator is no longer twelve since you arranged your chocolate into four equal groups called fourths.

7. Finally, arrange the twelve sections of your chocolate bar into three equal groups. Did you put four sections into each group? If you ate one of those groups, what fraction names the part that you just ate? (Answer: 1/3 or one-third) Which number is the numerator? (1) Which number is the denominator? (3)

The next time you're camping with your family and friends, take along the marshmallows, chocolate bars, and graham crackers. Make S'mores and enjoy some chocolaty fraction fun!

Geometry

□ Attribute Blocks:

These inviting little geometric shapes can readily be found in most primary grade-level classrooms. Attributes are qualities or characteristics about a person or thing. Therefore, attribute blocks come in varying colors, sizes, shapes, and thickness.

Students will be asked to sort and classify the shapes by arranging them together. You will need to provide parents and students with an activity guide something like this:

SORTING and CLASSIFYING

Dear Parents,

Developing logical thinking skills will foster healthy decision making for students now and in the future. Help your child to pour out the tub of desktop attribute blocks, and then sort and classify the blocks by the following:

      1. Same Color (yellow, blue, or red)

      2. Same Shape (square, circle, rectangle, hexagon, and triangle)

      3. Same Size (small or large)

      4. Same Thickness (thick or thin)

When you're finished with all four classifications of the blocks, you will earn a prize. Way to go!

    ▫ Pattern Blocks/Pattern Cards:

Like attribute blocks, pattern blocks are also readily found in most primary grade-level classrooms. Pattern cards, however, may need to be ordered through a catalog or purchased at a school supply store. Students will discover, without any necessary directions, that they will be matching the geometric size and shape of the pattern blocks with the images on the cards. This requires Spatial Reasoning, a concept that needs to be nurtured and developed. With proper development, some of us are able to visualize the rearrangement of a room full of furniture. We'll also know whether or not our vehicle will fit into that tight parking space! It wouldn't hurt to mention that in your posted activity guide for parents.

    ▫ 2-Dimensional Watercolor Paintings:

By first grade, most students are well-aware of four common 2-dimensional shapes: circle, square, triangle, and rectangle. By 2nd grade and beyond, this is expanded to include less-common shapes: pentagon, hexagon, octagon, and trapezoid.

Using bags of purchased plastic shape tracers (or you can make your own out of tag board), students will be tracing those eight shapes onto a piece of white 12x18" watercolor paper. They will enjoy painting the shapes with watercolor paints. Older students can overlap the shapes and then mix colors only in the areas where the shapes overlap. An adult should outline the painted shapes with a permanent black marker after the paint has dried. Students will visibly be attaching meaning to these shapes through the piece of art they are creating in the process.

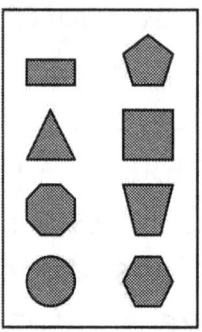

Among photographs, these watercolor projects could be displayed after hosting your Nachos and Numbers event.

- Texas-Size Area and Perimeter:
  The tiled floor of any cafeteria is an ideal way to teach the concepts of perimeter and area. Simply use masking tape to outline a large square, a rectangle, a right triangle ½ the size of your square, and any other shapes you have room for. Most floor tiles measure 1 foot by 1 foot, however if yours don't, you can count each tile or tile edge as a UNIT. Providing a guide, students will be asked to measure the perimeter (or outside rim) of each

shape, and then they'll be asked to measure the area (or inside space) of each shape. Students will discover that a square's perimeter is equal to its area, the area of the triangle is half of the square, as well as other memorable discoveries.

❑ <u>Don't Shoot My Attribute!:</u>

I personally wrote this silly little poem as a motivational, yet educational tool to teach fourth graders to respect and appreciate each others differences. It was then a simple transition to associate their own attributes with the attributes of 3-dimensional shapes (shapes with length, width, and height).

<u>Stop, Don't Shoot</u>!
Please don't shoot my attribute.
My mother really thinks I'm cute.
I may be short, and a little fat,
but my girlfriend thinks that I'm 'all that'!
I brush my teeth, and comb my hair.
I even wear clean underwear.
Friendly, helpful, and sometimes wise...
Let's cut this all right down to size.
My attributes are not all bad.
Of this I'm sure and aren't you glad?

Sadly, students in elementary classrooms often learn the names of 3-d shapes such as a sphere, prism, or pyramid by looking at pictures in a textbook, or on a worksheet. The attributes of those shapes determine their names. Providing sets of 3-dimensional shapes, introduce the attributes of faces, edges, and vertices by this suggested method. With a blindfold on, the student will name the number of faces, edges, and vertices for each shape before naming the shape. Then it's time for Mom or Dad to give it a try!

❑ Geoboards/Geoboard Activity Cards

What child could resist geoboards? What adult could resist geoboards? Again, there is likely a set or sets of these collecting dust on some teacher's shelves or in a closet of your school. Why? Geoboards are a wonderful way for students to create and discover geometry concepts, however they require guidance or questioning that many classroom teachers are not comfortable with. Also, the images of rubber bands flying through the air of classrooms perhaps frightens others. Therefore, I am suggesting that you purchase a boxed set of activity cards from any teacher's supply store that will lead students through their discoveries. It would be very appropriate to ask a student and their parent to complete five activity cards to complete this segment of Geometry, but they'll probably want to do more.

Statistics and Probability

❑ The Very Numerical Newspaper:

In this age of technology, the newspaper often gets over-looked as an outstanding educational resource, especially in the area of Mathematics. Using any city's newspaper, provide a questionnaire that requires participants to use the newspaper's index to answer questions about statistics. They'll also be learning a great deal about that day's news with questions similar to these:

1. Using the newspaper's index, find the last page

of the Sports Section where you will see the weather forecast. What was the high temperature predicted for yesterday? _____ degrees

2. Find out how much warmer the forecasted high temperature in Houston was than the high temperature in Boston on Sunday, February 9th by calculating the difference?
_____ degrees

3. With the Stock Show and Rodeo about to begin, you'll need a new pair of boots found on page 7B. What is the cost of
those boots? $_____

4. Do you own any shares of stock in a company? Using the Nasdaq listings on page 4C, what was the closing price of EbayInc stock? _____

5. On the front page of Section C, you will find an article about rising gasoline costs. How much has the price of a gallon of gas risen since a year ago? _____cents

6. In section A, you'll find a wonderful photo of an elderly Nebraska couple who just got married. What is the sum of their ages? _____years

7. The daily Houston Chronicle (or whatever newspaper you're currently reading) costs just _____cents.

❑ <u>Strange, But Interesting Facts With Numbers:</u>
Early on in my teaching career, I discovered that elementary-age students *love* going home at the end of each day with some trivial piece of knowledge that they can share with their parents. Those trivial pieces of knowledge are actual statistics that someone gathered through study and research. This activity integrates Mathematics with Reading comprehension, while providing participants with

trivial facts they can impress their parents and friends with. Post the following facts on laminated tag board, with quiz questions to follow. Of course they can look at the answers as they're completing the questions.

1. There are 293 ways to make change for $1.00.
2. The average person's left hand does 56% of the work on a computer keyboard.
3. The shark is the only fish that can blink with its 2 eyes.
4. There are more chickens in the world than there are people.
5. 2/3 of the world's eggplant is grown in New Jersey.
6. The longest one-syllable word in the English language is SCREECHED.
7. All 50 states are listed across the top of the Lincoln Memorial on the back of the $5.00 bill.
8. Maine is the only state whose name has just 1 syllable.
9. There are only 4 words in the English language which end in the suffix "dous": tremendous, horrendous, stupendous, and hazardous.
10. A cat has 32 muscles in its ear.
11. An ostrich's eye is larger than its brain.
12. In most advertisements, the time displayed on a watch is ten minutes after ten, or 10:10.
13. A dragonfly has a life span of only 24 hours.
14. A goldfish has a memory span of 3 seconds.
15. A dime has 118 ridges on its edge.
16. The average person falls to sleep in 7 minutes.
17. The giant squid, which is up to 200 feet long, has the largest eyes in the world.
18. There are 336 dimples on a regulation golf ball.

19. STEWARDESSES in the longest word in the English language that is typed only with the left hand.
20. In the state of Utah, when a person reaches the age of 50, then he or she can marry their cousin.
21. A mole can dig a tunnel 300 ft. long in just one night.
22. Over 10,000 birds a year die from smashing into windows.
    Have you watched Alfred Hitchcock's movie, <u>The Birds</u>, lately?

<u>Measurement</u>

❑ <u>The Awesome Angle Adventure:</u>
Household mini-blinds have become so affordable that they almost aren't worth cleaning anymore. Take them down and replace them once a year. The vinyl slats of those old mini-blinds can be cut with a scissors to make wonderful little tools for measuring acute (greater than 0 degrees, but less than 90 degrees), right (90 degrees), and obtuse (more than 90 degrees but less than 180 degrees) angles. Why not call it an AMT, or angle-measuring tool? Not only will students take pride in making their own, then you'll send them on an adventure through your Nachos and Numbers Event to measure angles. Have plenty of slats precut in equal lengths, with brass brads to fasten them together. Holes can be punches with a regular paper hole punch.

An angle is formed where two lines, line segments, or rays share a common point. There are three basic types of angles: right angles (form a perfect L or 90 degrees), acute angles (smaller than an L, less than 90 degrees), and obtuse angles (larger than an L, more than 90 degrees).

1. With your AMT, or angle measuring tool in hand, find a table with base ten flats (or 100's) on it. Does the corner of a flat form an acute, right, or obtuse angle?

2. Find another table with moveable school clocks on it. Set one of those clocks to read 6:15. Using your AMT, do the hands of that clock form an obtuse, acute, or right angle?

3. For your final adventure, find any man willing to spread their arms out in front of them in a V-shaped pattern. Where ONE of their arms meets their chest, is that an acute, right, or obtuse angle?

❑ <u>Symmetrical Tempra Paintings:</u>
  I find it very sad that many school districts today

do not provide Art teachers at the elementary school level. There is so much spatial reasoning taught through quality art instruction, as well as other Mathematics skills. This simple project will introduce young students to symmetry while allowing them to be creative. Provide students with 9 x 12" white construction paper and three or four primary colors of tempra paint complete with brushes. They'll first be told to fold the paper in half either horizontally or vertically. Open the paper back up, and they can then paint anything simple that they want on only one side of that folded piece of construction paper, applying the paint liberally. When they're finished and before the paint is dry, ask them to carefully fold the paper closed and press down firmly. When they open it back up again, the image painted on one side of the paper will now be transferred to the other side as well. In other words, the two images are mirrored reflections, or symmetrical. These projects can also be prominently displayed after your Nachos and Numbers event!

❑ <u>Symmetry as EZ as A-B-C:</u>
Developing the concept of symmetry allows students to visualize the relationship of equivalence. Provide students with a word-processed copy of all sixteen letters of NACHOS AND NUMBERS in basic block form.

NACHOS AND NUMBERS

Using a ruler, students will be asked to draw any and <u>all</u> lines of symmetry that perfectly divide that letter into two equivalent parts. Here is an example of a letter of the alphabet displaying its lines of symmetry

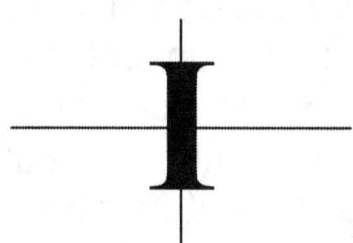

- <u>Measurement Madness:</u>

As teachers and parents, we can talk about measurement until we're "blue in the face". In reality, students have to physically measure things in order for units of measure to have real meaning. It shouldn't be done once a year during a textbook chapter about measuring either. Measuring needs to take place often and meaningfully. These little measuring tasks are meant to be fun, yet meaningful. If a participant remembers just one small part of this activity, then you've succeeded as a facilitator.

In perhaps gallon-size plastic bags, provide participants with items such as the following. Make certain that the items are the same size and length. Students and their parents will be asked to measure the length, width, diameter, or the circumference of the contents of each item in the bags.

a. a 2 ft. piece of Weed Cutting Line (length)

b. a Compact Disc, like the ones we often receive in the mail from internet providers, (diameter, circumference or both?)

c. an unsharpened new pencil (length)

d. a Popsicle stick (length and width)

e. a 3-inch long metal bolt (length)

f. a brightly-colored index card (length and width)

g. plastic coins such as quarters (diameter)

Providing students with both rulers and tape measures, ask them to measure these items in both standard and metric units. They will quickly be able to determine which measuring tools will be most effective in completing the assigned task.

Addressing Multiple Strands

As mentioned earlier, there are activities you can set up for your Nachos and Numbers Event that will address several strands of the Mathematics curriculum at the same time. More than one of these activities involve technology, vital to a student's success in today's world.

❑ The Mini-Cyber Mathematics Café:

Cyber Cafes have popped up in communities all over the world. People can enjoy a cup of coffee while surfing the net, checking their e-mail, or just enjoying the fellowship of friends getting together. You can achieve the same results during your Nachos and Numbers event on a small scale. Simply arrange 4-6 personal computers or laptops together, loaded with Mathematics software already used in your school's technology lab or within individual classrooms. You must project one of those PC's onto a large screen. This will attract participants from all over your event, especially software as motivating as Turbo Math. What's Turbo Math, you ask? After completing several tasks where players earn cash, players get to purchase a race car and then race around a track.

Though there is a ton of excellent Math software out there, these are a few of my own personal favorites popular with my own students, yet applicable to a variety of elementary and middle-school skill levels:

1. Turbo Math
2. Clock Works
3. The Math Majors (baseball)
4. M & M's, The Lost Formulas
5. Wrangler Math
6. Coin Critters
7. Turbo Math Facts
8. The Kidville Series of software

I am also going to suggest that you provide parents with a list of websites that will provide their children with free Math games and activities they can access from their household personal computer. Here are just a few excellent sites that are easily accessible:

www.math.com

www.kidsdomain.com

www.coolmath4kids.com

www.gomath.com

www.kidskonnect.com

www.aplusmath.com

www.funbrain.com

www.primarygames.com

- Calculators/Calculator Activity Cards:
  These very affordable sets of boxed cards are a motivating way for students to achieve comfort and skill in using calculators.

  Any simple calculator will do, even the dollar-store variety. The particular boxed set of cards that I like get increasingly difficult with chronological number, so you can easily assemble cards in groups of 10 within zip-lock bags to meet the needs of a variety of age and ability levels. When a student completes a short series of operations on the calculator, he or she is asked to turn the calculator over to reveal the answer to the riddle at the top of the card. They're fun, yet highly engaging for both kids and adults.

- The Math Literature Lounge:
  Combining Literature with Mathematics units has become very popular as educators have discovered the value of integrating subjects vs. teaching skills in isolation. Public school librarians have done a wonderful job of ordering books, often at teachers' requests, to meet the growing popularity. Arrange several bean bag chairs in an isolated corner of your Nachos and Numbers Event. Or, for a quieter

environment, set the Lounge up in a room within close proximity of your event. Provide reading lamps, rugs, or even large pillows to enhance the attraction. Display on shelves that separate the "lounge" from the rest of the event perhaps two dozen of your favorite pieces of Math literature. What parent could resist reading one of these great books to their child? This idea will introduce parents to the wide variety of Mathematics Literature that can be checked out of your school's library. Here is a very condensed list of popular Mathematics literature available today with a variety of topics:

The Big Buck Adventure
Big Numbers
Counting Is For The Birds
Domino Addition
Grandfather Tang's Story
How Did Numbers Begin
Herbert Hilligan's Prehistoric Adventure
Herbert Hilligan's Lone Star Adventure
Herbert Hilligan's Tropical Adventure
Herbert Hilligan and His Magical Lunchbox
If You Made a Million
The Greedy Triangle
The Icky Bug Counting Book
Sir Cumference and The Great Knight of Angleland
Skittles Riddles
Subtraction Action

Any elementary school library should consider purchasing a book published by NCTM, The National Council of Teachers of Mathematics, that directly correlates literature that will enhance the National Math Standards. It is available at this website: www.nctm.org/eresources

Chapter 9

Rewarding Participation

Though sometimes controversial among educators, rewards and incentives for attending an event such as Nachos and Numbers are important to the success of hosting annual events that attract large numbers of participants. For children like my own, I've learned that it has to be more than just F-U-N.

They expect to end the evening carrying home the rewards of their efforts and achievements.

Hopefully, you achieved enough monetary support in the beginning of this project to avoid any of your own out-of-pocket expenses. I mentioned ideas for that support in Chapter 3, in case you didn't.

Prizes earned through participating in large- floor activities like those addressing Basic Facts should be immediate: cupcakes and ½ liter sodas are already earned during two different activities. However, other large-floor activities would best be served by prizes given out immediately upon completing that activity, such as: frizbees, kites, yard/meter sticks, and coupons to area restaurants. By the way, reading a good book in the Math Literature Lounge should be rewarded with something children can't resist. After all, good Math teachers need to be good Reading teachers, right?

I've personally found it easiest to award **tickets** for completed table activities. Tickets can be purchased in large rolls very inexpensively. Activity signs displayed can also explain to a participant how many tickets can be earned for completing that activity. As a student and their parent complete a table activity or game, they should approach a member of the <u>Prize Patrol,</u> roaming throughout your scheduled event. Wearing brightly-colored vests or aprons, they should be highly visible as well as friendly. It is their job to award tickets to students accompanied by a parent, on the honor system. Those tickets can then be "cashed in" for prizes from the <u>Prize Princess.</u> Any smiling, new teacher would be honored to wear a rhinestone-jeweled tiara, seated at this table of honor.

The prizes awarded by the <u>Prize Princess</u> are the kind ordered in large quantity through a catalog such as The Oriental Trading Company, mentioned in Chapter 4. Ideas for those include: sunglasses, fancy pens and pencils, nail polish for girls, sun visors, hats, necklaces and bracelets, toy cars and trucks, rubber bouncing balls, fancy drinking straws, dice, etc. You'll see multitudes of ideas as you page through any catalog. Simply call 1-800-228-2269 to receive your own catalog.

The rewards for parents attending the Nachos and Numbers Event are somewhat intrinsic or heart-felt: a free meal, a good time had by their family, as well as an afternoon or evening spent doing quality educational activities with their children.

As the facilitator for such an event, the best reward I've received is hearing this question from parents and students as they've departed, **"Are we going to have Nachos and Numbers again next year? This was lots of fun!"**

Chapter 10

Enlisting Volunteers to Help You

It might be just a couple of weeks before your scheduled Mathematics event. The activities are prepared and organized by strand, perhaps in boxes or baskets ready for set-up and display. You can't possibly be everywhere you need to be to ensure that everything flows smoothly. Therefore, it's time to garner support beyond the colleagues or parents who've helped you out thus far. Here are some suggestions for seeking the kind of volunteers who would likely be honored that you would ask for their assistance.

1. College/University Education Students
2. Senior Volunteers (especially if you already have a mentoring program in place)

3. Area High School Athletes/Cheerleaders/Band Members

4. Former Students, now at the Middle or High School Level

5. Your School's Administrators, as well as your District's Administrators

6. School Board Members

7. Teachers (volunteers only)

I find it necessary to mention why I only suggest using classroom teachers as a workforce if they volunteer. Students spend every day of the school year with their classroom teachers. Idealistically, they likely already know that those individuals value Mathematics. It is important for students to realize that people from all walks of life value Mathematics, and that it can be lots of fun. Your school's classroom teachers, with children of their own, can therefore enjoy the evening themselves as a family!

Since Nachos and Numbers is only scheduled for 2 ½ hours in length, you can likely get by with as few as eight to ten volunteers. They'll need a short briefing before the event begins so they completely understand their tasks, whether they're serving food, serving on the prize patrol, monitoring a specific activity such as Multiplication Bingo, or even roaming the event floor taking candid photographs. Don't forget to have friendly folks greeting students and parents at the door. I also suggest that you compose and photocopy a written greeting such as this explaining the logistics of your event. Include a reverse side of the handout in Spanish of course.

Dear Parents and Students,

<u>Welcome to Nachos and Numbers</u>! Nachos Grande' and refreshments are being served throughout the evening for your enjoyment. You may eat anytime throughout the event.

The cafeteria is set up with table activities as well as large floor activities. They are all labeled for your convenience, but feel free to ask any questions that you might have. Large floor activities are rewarded with immediate prizes, while table activities are rewarded with tickets from the roaming Prize Patrol. Those tickets can be redeemed for prizes from the Prize Princess, seated behind Prize Central. For your enjoyment, Bingo will be played throughout the evening as well.

We want to thank you for attending Nachos and Numbers.

For safety reasons, we ask that you stay with your children at all times. We sincerely hope you enjoy an evening of festive Mathematics fun! Please take time to fill out a short questionnaire before you leave. We value your comments.
    Gracias!

My wife and I have had lots of fun collecting sombreros, and brightly-colored aprons for our own Nachos and Numbers events. This way, we can easily make every effort to make sure our volunteers are adorned in festive attire! There are more ideas in the next chapter. These generous volunteers will also need to be well-fed throughout the evening, and having fun themselves. It'll make a substantial difference in the success of your entire family Math event if everyone is happy and feeling appreciated.

Chapter 11

Staging a Festive Event

Staging the presentation of any festive occasion can be lots of fun. The South-of-the-Border theme for Nachos and Numbers lends itself well to a variety of ideas. Here are some of my own.

Mariachi music has a rich history all its own. Who could resist the magical blend of trumpets, guitars, and Latino voices enriching an occasion such as Nachos and Numbers? Set up a quality CD player, or play music through your school's PA system. I am going to recommend two particular groups for your choosing, both with websites that are easily accessible for ordering compact disks. One is called Mariachi Vasquez, based in Denver, Colorado. Their website is as follows:

www.mariachi@gte.net

The other is called Mariachi USA, based in Hollywood, California. Their website is as follows:

www.mariachiusa.org

Mariachi Vasquez is a well-known award-winning family of musicians that produces irresistible music. Mariachi USA is known for their philanthropic support for music in schools. Now that you've discovered wonderful "sounds"… let's focus on the "sights".

You can inexpensively cover tables in vinyl tablecloths of red, orange, yellow, and green. Ask any creative teacher to have his or her students make bouquets of tissue paper flowers on pipe cleaner stems to decorate the tables as well. Wouldn't papier mache' piñatas made and painted by students be an awesome addition to your event?

Does someone you know have strands of chili pepper lights used only at Christmas? If not, they're available any time of the year through catalogs like the Oriental Trading Company. Worth the minimal investment, you'll use them year after year.

Prize Patrol volunteers should be wearing brightly colored aprons with pockets for their tickets. Food Servers should be wearing aprons and sombreros as well! While on the subject of food, one of my own favorite "finds" in seeking out fun, festive serving pieces was to discover small yellow plastic sombreros that once held salt for salting the rim of a margarita glass. They make wonderful spoon rests for the garnishes served with your nachos. They could also be used to hold dice for any table activity, such as Roll an Array, described in Chapter 8.

Helium balloons tied to tables will give any event a feeling of celebration. Why not use those balloons as rewards for a specific activity such as the Basic Facts Bean Bag Toss?

Hopefully, I've given you a few ideas that will motivate your own creative thinking for staging a festive Nachos and Numbers event. Though seemingly unimportant, the smallest detail can make an integral difference in the success of any occasion. Having fun is what it's all about, and your efforts are bound to bring smiles to the faces of children and parents in attendance.

Chapter 12

Be the Facilitator that will

<u>Make This Happen!</u>

As busy, sometimes overwhelmed educators, it is often difficult to imagine taking on the additional work of another project that will require hours beyond the regular work day. My own experience of twenty-three years has taught me, however, that when I truly believe in the educational benefits of an idea or proposal, the time and energy necessary to facilitate it "magically flow". It's an energizing feeling! Not only will that idea become an accomplished reality, I'm often astonished at how much fun I had in the process.

You, therefore, have the opportunity to become the driving force behind implementing your own Family Math Event. Start months in advance, negotiate solid administrative support, secure funding and resources, and then get busy.

In conclusion, remember to provide attendees with a short and simple survey to provide you with feedback for program evaluation purposes before exiting the evening festivities. Those survey questions can be as simple as these:

1. What was your favorite activity during our Nachos and Numbers Family Math Event?

2. Did the event help you to better understand some of the Math concepts being taught in the classroom?

3. Can you give us any suggestions as to how we can improve Nachos and Numbers for next year?

As you well know, there are distinct rewards for going "above and beyond" the call of duty, no matter what your occupation may be. They may <u>not</u> be the kind of rewards you can see and touch. However, months and even years later, you'll continue to experience the intrinsic rewards of your hard work when you are fondly remembered by former students and their parents. Your achievements will be celebrated. Your hard work, enthusiasm and dedication will have made a difference in people's lives. That is a call for celebration in itself!

I sincerely hope your Nachos and Numbers Family Math Event is wonderful success.

About the Author

Thomas John Ecker is a classroom teacher with twenty-three years of experience teaching Mathematics. That experience includes teaching all levels of elementary education, middle school, as well as serving in the capacity of Elementary Mathematics Coordinator.

Mr. Ecker's accomplishments and achievements as an educator have earned him several education honors, including Outstanding Young Educator, Education Grant Recipient, as well as Teacher of The Year. In April of 2004, he was a featured speaker at the National Council of Teachers of Mathematics (NCTM) Conference in Philadelphia, Pennsylvania.

Mr. Ecker is married to another dedicated teacher, Jaynie, and together they have five children.

Those great kids, and hundreds of others, have been the inspiration for the creative and very practical ideas within this guide. Let it be the source of your own inspiration in hosting a successful family Math event, NACHOS AND NUMBERS!